THE

POWER

OF

ONE

1

Dr. Marcia R. Pinheiro

-

Prologue

Hollywoodians had a dream: Six degrees of separation would be the maximum we had in terms of communication in human kind.

Their dreams were well fed by the media: In Brazil, everyone was paying attention to Duncan Watts and what the press members said, not necessarily the own Duncan, about his work.

It was said that anyone can reach anyone else on earth in at most six communication moves, basically: We get the most oppressed member of society, so say a cleaner, and ask him to find a way to speak to the English queen in person. He should definitely be able to get that done, for in at most six steps everyone can reach anyone else on earth.

They then believed the power of one or the power of many? With one person who is intelligent, our message could get, with no mistake, to any other human being in the Globe. Notwithstanding, for that to happen, and to justify our claim that this person would have to be intelligent, a few other people needed to be used. Oh, well, the theory said at most six steps, not exactly six always: Some people could get to the other in one step, obviously and trivially.

The Six Degree Theory was then the expression of a human dream: We can always reach the person we want to reach; it is just a matter of strategy.

In 2002 we were obliged to do some serious research into Comellas et al's findings, and that led to our studies on the Six Degree Theory.

It was only years later, however, thanks to a lot of crimes suffered, that we reached the words of the psychologist,

Milgram: He was the actual creator of the Six Degree Thesis.

Unfortunately, the own Milgram had proven that thesis to be false: Even with his letters, and therefore without any computer or serious mathematical calculations, he had been able to prove that the thesis couldn't possibly be true.

We later on got serious reassurance on his conclusions being correct. That appeared in our paper, Starants II.

Even though this dream, of always being able to reach the person we want to reach, has been consistently proven to be nothing beyond that, a dream, we can still find plenty of reasons to believe in the power of one, and plenty of reasons not to believe in the power of many instead.

With this book, we intend to develop a scientific theory, but this theory cannot be totally mathematical or logical, since it is a theory involving human beings, and human beings who are alive. This theory has to be statistical or contain a few examples with weight that is superior to that of the other examples that could be presented as counter-argument to our thesis.

Introduction

Mata Hari is one of the most important feminine figures of all times: She is told to have disgraced entire countries because all men were seduced and commanded by her.

Margaret Tatcher is in the same category: She graciously commanded several men and men are told to have obeyed her in all that she said.

One man pushed the button that destroyed Nagasaki. One man pushed the button that destroyed Hiroshima.

One man made the Japanese lose through following whatever instructions the CIA gave him, but that far the Japanese, his people, were about to become first in place of the USA.

One man changed human history by shooting JFK.

JFK is told to have caused Monroe's suicide. Again one man.

Because of Princess Isabel the slaves found their freedom in Brazil by 1800: One woman.

Because of Levi Strauss, we nowadays have strong pants, which, in some cases, last forever, like considering the length of our individual lives.

Because of Steven Spielberg, we started thinking that the deformed bodies found in special storage in the USA could contain even adorable beings.

The power has to be on one.

We will now detail some of the stories we just mentioned on the intentions of making the reader notice why the power is definitely on one, not on many.

ET the Extraterrestrial is a movie from 1982. According to the IMDB, it was written by Melissa Mathison.

This is the storyline from IMDB:

After a gentle alien becomes stranded on Earth, the being is discovered and befriended by a young boy named Elliott. Bringing the extraterrestrial into his suburban California house, Elliott introduces E.T., as the alien is dubbed, to his brother and his little sister, Gertie, and the children decide to keep its existence a secret. Soon, however, E.T. falls ill, resulting in government intervention and a dire situation for both Elliott and the alien.

And the financial details:

Budget:

$10,500,000 (estimated)

Opening Weekend:

$11,911,430 (USA) (11 June 1982)

Gross:

$434,949,459 (USA)

Spielberg's ET

This is the picture of an ET that they claim the Pentagon knows about:

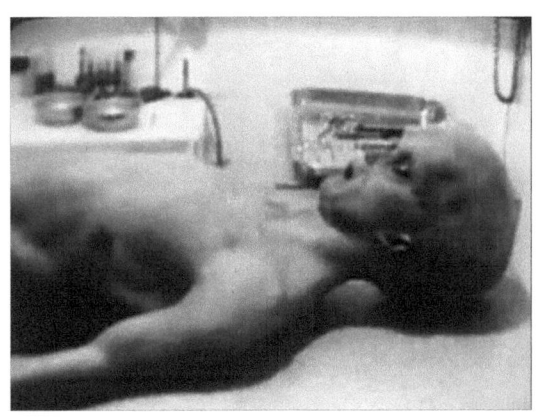

And that is why we are now keener on literally hugging and kissing the ETs.

That is likely to be a dangerous activity, yet our lives are changed forever and there will certainly be lots of people willing to give love to the creatures if they ever come.

You see, in our World of Fantasy, Spielberg could be an alien himself, disguised as human, and he could then wish for creating good will because the creatures are weaker, exist in smaller numbers, and need our life energy to exist, for instance.

We approach them full of love and they basically suck our hearts into their bodies to keep on living forever, let's say.

The nature of human kind is obviously hostile, so that we

would normally tend to attack the creatures.

Spielberg could be the person who made us all die whilst believing we have deep love from the creatures, basically.

Melissa did write it, but, as we know, we could utter that it is almost scientifically correct to say that, without Spielberg, nobody would really love ETs.

The power of one that may become the failure of many, basically.

Mata Hari was born in the Netherlands according to

http://www.biography.com/peo
ple/mata-hari-9402348

She was a spy and she spied for France according to the same source.

Even though her spying activities are legendary for those who were in this world back then or who were just a few years away from those, such as Murillo Francisco Ribeiro de Assis Puget, the best source (http://www.firstworldwar.com /bio/matahari.htm) we got said the following:

Still unclear today are the circumstances around her alleged spying activities. It was said that while in The Hague in 1916 she was offered cash by a German consul for information obtained on her next visit to France. Indeed, Mata Hari admitted she had passed old, outdated information to a German intelligence officer when later interrogated by the French intelligence service.

Mata Hari herself claimed she had been paid to act as a French spy in Belgium (then occupied by German forces), although she had neglected to inform her French spymasters of her prior arrangement with the German consul. She was, it seemed, a double agent, if a not very successful one.

It appears (the details are vague) that British intelligence picked up details of Mata Hari's arrangements with the German consul and passed these to their French counterparts.

She was consequently arrested by the French on 13 February 1917 in Paris. Following imprisonment she was tried by a military court on 24-25 July 1917 and sentenced to death by a firing squad. The sentence was carried out on 15 October 1917 in Vincennes near Paris. She was 41.

This is Mata Hari according to https://www.1001freedownloads.com/free-clipart/mata-hari

This is her too, this time according to https://au.pinterest.com/pin/374361787755140093/

https://www.reference.com/history/won-world-war-eceef70d9766490c tells us that the WWI went from 1914 to 1918 and France was one of the winners.

She was arrested in 1917 according to the sources we here mention, so that she could easily have been responsible for France being one of the winners, as plenty, such as Murillo, may say.

The power of one woman: To give the war to a certain Country and to take it from others, and therefore the power of one that may become the failure of plenty as well as the success of plenty.

Who was the man who pushed the button that determined the fate of Hiroshima and Nagasaki?

http://www.history.com/topics/world-war-ii/bombing-of-hiroshima-and-nagasaki lets us know the following:

On August 6, 1945, during World War II (1939-45), an American B-29 bomber dropped the world's first deployed atomic bomb over the Japanese city of Hiroshima. The explosion wiped out 90 percent of the city and immediately killed 80,000 people; tens of thousands more would later die of radiation exposure. Three days later, a second B-29 dropped another A-bomb on Nagasaki, killing an estimated 40,000 people. Japan's Emperor Hirohito announced his country's unconditional surrender in World War II in a radio address on August 15, citing the devastating power of "a new and most cruel bomb."

Hiroshima, a manufacturing center of some 350,000 people located about 500 miles from Tokyo, was selected as the first target. After arriving at the U.S. base on the Pacific island of Tinian, the more than 9,000-pound uranium-235 bomb was loaded aboard a modified B-29 bomber christened *Enola Gay* (after the mother of its pilot, Colonel Paul Tibbets). The plane dropped the bomb–known as "Little Boy"–by parachute at 8:15 in the morning, and it exploded 2,000 feet above Hiroshima in a blast equal to 12-15,000 tons of TNT, destroying five square miles of the city.

Hiroshima's devastation failed to elicit immediate Japanese surrender, however, and on August 9 Major Charles Sweeney flew another B-29 bomber, *Bockscar*, from Tinian. Thick clouds over the primary target, the city of Kokura, drove Sweeney to a secondary target, Nagasaki, where the plutonium bomb "Fat Man" was dropped at 11:02 that morning. More powerful than the one used at Hiroshima, the bomb weighed nearly 10,000 pounds and was built to produce a 22-kiloton blast. The topography of Nagasaki, which was nestled in narrow valleys between mountains, reduced the bomb's effect, limiting the destruction to 2.6 square miles.

This is what we have heard so far in life: How horrible it all was for the Japanese.

Now let's see what they have to say in favour of these two men who basically sealed the destiny of these 120,000 Japanese people, mostly civilians, on top of all the side effects that would appear in all other people around those for perhaps the entire term of their lives:

By the time of the Trinity test, the Allied powers had already defeated Germany in Europe. Japan, however, vowed to fight to the bitter end in the Pacific, despite clear indications (as early as 1944) that they had little chance of winning. In fact, between mid-April 1945 (when President Harry Truman took office) and mid-July, Japanese forces inflicted Allied casualties totaling nearly half those suffered in three full years of war in the Pacific, proving that Japan had become even more deadly when faced with defeat. In late July, Japan's militarist government rejected the Allied demand for surrender put forth in the Potsdam Declaration, which threatened the Japanese with "prompt and utter destruction" if they refused.

General Douglas MacArthur and other top military commanders favored continuing the conventional bombing of Japan already in effect and following up with a massive invasion, codenamed "Operation Downfall." They advised Truman that such an invasion would result in U.S. casualties of up to 1 million. In order to avoid such a high casualty rate, Truman decided–over the moral reservations of Secretary of War Henry Stimson, General Dwight Eisenhower and a number of the Manhattan Project scientists–to use the atomic bomb in the hopes of bringing the war to a quick end. Proponents of the A-bomb–such as James Byrnes, Truman's secretary of state–believed that its devastating power would not only end the war, but also put the U.S. in a dominant position to determine the course of the postwar world.

All we know of is that tons of things could have been done, including lacrimal gas, sound waves, and others, tons of things that would not cause the death of civilians.

In this case, who caused all those human losses, of civilians, was Truman, the American president.

If they consulted academics, like myself, it would be likely that we would come up with much better strategies in order to save both the one million lives of the US and the hundreds of thousands of lives of the Japanese people.

The power of one man: Finishing with the unique opportunity of hundreds of thousands of people on earth gratuitously, and therefore the power of one that becomes the power of God, that of giving or taking the rights to exist in human shape, the rights to live.

Who killed JFK was the former US marine Lee Harvey Oswald, according to the dominant media.

http://www.mirror.co.uk/news/world-news/oliver-stone-reveals-who-really-8729805 brings another version of the facts, however. See:

"Everything would go to a PO Box, and he'd locate it from offshore, from Bermuda."

Then detailing the moment they met in a hotel in Rochester, New York, Stone recalled: "He said he didn't want money or recognition. He said something like, 'I want you to know this is from my conscience'.

"The scenario he laid out was very practical. It's the way I would do it, if [I] were going to do something like that.

"You kill the President, and your cover is security, and if the sniper or snipers who kill the President are hidden in with the guys who are supposed to protect him, guys who have no knowledge of this plot... It makes a lot of sense.

As a former marine in Vietnam, the director said he was left convinced by the "military jargon" and intricate details within an account that he describes as "plausible" and "very authentic".

Stone revealed the man's confession for the first time to Matt Zoller Seitz, who is the author of a forthcoming book on the Oscar-winning screenwriter whose films include Platoon.

When asked why Stone had kept it secret so long Seitz said: "I think it was because he trusted me, and also because both the father and the son have been dead for a while.

"Nobody has ever heard this story. I'm the first person."

In his book, Seitz tells how Stone initially received Ron's messages.

The director told him: "He came to me through a series of weird letters through post office boxes.

If we could have access to the mentioned letters, it would be easier to judge, but, in any hypothesis, JFK would have been killed by shots that came from a single gun.

Since Lee ends up killed two days after they say he was

guilty, it is to imagine that he wasn't.

Had he been told to have committed suicide, that would have been more logical: Trying to escape the charges, the punishment, etc.

https://www.jfklibrary.org/JFK/Life-of-John-F-Kennedy.aspx brings unexpected details. See:

By June 11, 1963, however, President Kennedy decided that the time had come to take stronger action to help the civil rights struggle. He proposed a new Civil Rights bill to the Congress, and he went on television asking Americans to end racism. "One hundred years of delay have passed since President Lincoln freed the slaves, yet their heirs, their grandsons, are not fully free," he said. "This Nation was founded by men of many nations and backgrounds...[and] on the principle that all men are created equal." President Kennedy made it clear that all Americans, regardless of their skin color, should enjoy a good and happy life in the United States.

The same source brings some expected details as well:

> ## The President is Shot
>
> On November 21, 1963, President Kennedy flew to Texas to give several political speeches. The next day, as his car drove slowly past cheering crowds in Dallas, shots rang out. Kennedy was seriously wounded and died a short time later. Within a few hours of the shooting, police arrested Lee Harvey Oswald and charged him with the murder. On November 24, another man, Jack Ruby, shot and killed Oswald, thus silencing the only person who could have offered more information about this tragic event. The Warren Commission was organized to investigate the assassination and to clarify the many questions which remained.

This source seems to wish for implying that JFK would have been killed because he was defending the end of racism.

We observe that the extract refers to occurrences from June, 1963, but the president was killed in November, 1963, according to the same source.

We hear nothing about JFK referring to the civil rights issue in the media in a meaningful manner after that event, from June, 1963.

If we really want to find something that is a fact that could connect to his assassination, perhaps we should look for events that are closer to the date of his death.

That is just a choice, speculation, it seems.

If the marine kills JFK, the reasons are likely to be military ones.

If the security guard kills him, the reasons could be personal.

If the guard acts in the name of a group, then it could not be the group of the security guards, could it?

If the marine acts in the name of a group, it is pretty weird that the conclusion is that he acted in the name of the marines, since those should stick together.

If they act as individuals, however, the cause could as well be that the president caused the death of Marilyn Monroe, and that happened on August, 1962. http://www.history.com/this-day-in-history/marilyn-monroe-is-found-dead brings the details. See:

On August 5, 1962, movie actress Marilyn Monroe is found dead in her home in Los Angeles. She was discovered lying nude on her bed, face down, with a telephone in one hand. Empty bottles of pills, prescribed to treat her depression, were littered around the room. After a brief investigation, Los Angeles police concluded that her death was "caused by a self-administered overdose of sedative drugs and that the mode of death is probable suicide."

http://channel.nationalgeographic.com/killing-kennedy/articles/the-sex-life-of-jfk/ says the following:

- **Judith Campbell Exner:** The Los Angeles socialite, who had previously dated Frank Sinatra, began her affair with Kennedy during his 1960 Presidential campaign, according to her 1999 *Los Angeles Times* obituary. Exner and Kennedy continued their relationship during his term of office. During that time Campbell Exner claimed to have served as a courier for the President, taking mysterious envelopes to mobsters Sam Giancana— with whom she later also had an affair—and Johnny Rosselli. In 1963, at her final encounter with Kennedy, she became pregnant and later had an abortion, by her account. In 1975, her relationship with Kennedy became public when she was called to testify by a Senate committee investigating alleged CIA plots to assassinate Cuban leader Fidel Castro.

Now, we all get curious: When was her last encounter with JFK?

The other source said she died in 1962, so that we are getting really confused here.

See what http://www.smh.com.au/news/world/kennedy-link-to-death/2007/03/16/1173722744304.html says:

Critically, it raises an alleged conspiracy, apparently overseen by Lawford, for Monroe to unwittingly commit suicide with the drug Seconal, a barbiturate used to treat insomnia and relieve anxiety. The document gives no precise reason why she would be killed but hints it may be linked to her threats to make public her affair with Kennedy, as other conspiracy theories have previously claimed. It states in part: "Peter Lawford, [censored words blacked out] knew from Marilyn's friends that she often made suicide attempts and that she was inclined to fake a suicide attempt in order to arouse sympathy.

"Lawford is reported as having made 'special arrangements' with Marilyn's psychiatrist, Dr Ralph Greenson, from Beverley Hills. The psychiatrist was treating Marilyn for emotional problems and getting her off the use of barbiturates. On her last visit to him he prescribed Seconal tablets and gave her a prescription for 60 of them, which was unusual in quantity especially since he saw her frequently. On the date of her death … her housekeeper put the bottle of pills on the night table. It is reported that the housekeeper and Marilyn's personal secretary and press agent, Pat Newcomb, were co-operating in the plan to induce suicide."

FBI file links Kennedy to Monroe's death

March 17, 2007

Page 1 of 2 | Single page

For four decades there have been rumours that Marilyn Monroe's death was not a simple suicide. Now a Los Angeles-based Australian writer and director, Philippe Mora, has uncovered an FBI document that throws up a chilling new scenario.

BOBBY KENNEDY'S affair with the screen idol Marilyn Monroe has been documented, but a secret FBI file suggests the late US attorney-general was aware of - and perhaps even a participant in - a plan "to induce" her suicide.

The romance with Bobby Kennedy seems to be supported by, for instance, the words of his own wife (http://www.dailymail.co.uk/news/article-3811200/Marilyn-Monroe-DID-affair-Bobby-Kennedy-Newly-unearthed-letter-reveals-RFK-s-sister-said-pair-item-early-Sixties.html):

A letter from Jean Kennedy Smith (pictured) has given perhaps the clearest indication yet that the pair were together

Rumors over the pair's relationship started to fly after they were pictured together at a number of events - however, there was no concrete evidence at hand to confirm to they were an item.

Accounts from those present at JFK's 40th birthday party in 1962, at which the actress sung 'Happy Birthday,' say she went after Bobby - even in front of his wife Ethel.

'She literally pinned him against he wall, and she had him trapped,' claimed pollster Lou Harris.

'Ethel got so disgusted. When she got him home, she said, "That's the most disgusting thing I've ever seen."'

She sang Happy Birthday, Mr. President in May, 1962. See:

Fifty-four years later, one might suppose that the famous dress came on to the stage of its own accord, as if it had life as well as destiny. But yes, in 1962, there was a warm body inside it. This was 19 May 1962, and Marilyn Monroe was bringing happy birthday wishes to the president of the United States. She was 10 days early: the actual birthday of John F Kennedy, his 45th, would not be until 29 May. So what? The Democratic party wanted to have a super fundraiser at Madison Square Garden in New York City, so they needed not just JFK himself, but a hook and bait. The birthday was a pretext until the breathy song, "Happy birthday, Mr President ..."

That was Marilyn Monroe. But did the Dems ask her or did the president himself arrange it? Sometimes a president can be an instrument in his own PR. Marilyn would be 36 on 1 June, which would prove to be her last birthday. On 5 August, she was found dead in her bed in Brentwood with just a sheet around her.

So, if I find out that she has an affair in May, 1962, I could wish for killing her in August, 1962, or I could have asked people to plan things well and induce her to suicide, what could take that long to happen, given how famous she was, how much her employees liked her, etc. I certainly would kill her before I kill my husband.

However, if people convinced me that who had an affair with her was JFK, and I had killed her myself somehow, say with the help of those people, because I wrongly believed that she had an affair with my own husband, I would then kill JFK because he had used my good man for his filthy purposes, and that led me to do what I did.

http://www.dailymail.co.uk/news/article-2646535/Exclusive-I-spending-rest-life-Jack-Kennedy-Jackie-wanted-divorce-serial-cheating-husband-Marilyn-Monroe-straw.html says the following:

Trouble: Jackie was prepared to divorce Jack and become the only First Lady to divorce a sitting president, say the authors of a blockbuster new book about Jackie

Kennedy scion Joe Kennedy offered his daughter-in-law a check for one million dollars to stay in the union. Jackie allegedly told him that the cost of her staying with JFK would rise to twenty million dollars 'if he brings home any venereal disease from any of his sluts', according to the authors.

If Jackie had divorced Jack while he was a U.S. Senator, the old man's dreams of having a son in the White House would have gone unfulfilled. And he was more concerned about those political ambitions then he was about Jackie being ostracized by the Catholic Church.

So, now we have an angry wife because of the affair (Bobby's), an angry wife because of the affairs (JFK's), and serious political reasons (if he did not die, he would probably be the first president to get a divorce during his mandate. Worse, he was told to be Catholic).

We are told that things were getting progressively public, the thing about Bobby and Monroe more than anything else.

What they spread in the press back then was that who had an affair with Monroe was JFK, what could be a distraction, so say Bobby's wife plots against

Monroe and Bobby and/or the CIA people who liked him rush to put JFK on the spotlight instead, what makes Bobby's wife kill him next and give it all for settled.

A relative with high level of concerns in what regards morality, family name, and political perspectives due to propaganda could rush and kill JFK before the worst happens, so say a scandal that meant divorce. You know, whilst they are married, people may wonder, but will usually think that the rumours are not true. If they split, as it has happened, for instance, with Princess Diana and Prince Charles, people may start acquiring certainty, even because the woman will speak in case that was the truth.

In this case, the theory involving the security guard sounds way more plausible than the theory involving Lee, the marine.

The events surrounding JFK's life by that time seem to have been unimportant in terms of politics. He is told to have met an important political figure of the USSR back then, in October, as we can read on http://jfk14thday.com/jfk-gromyko-cold-war-1963/

That was on the 18th, October, 1963, and this is then the event that would be most likely to cause a death by marine's hands.

November, 21st DOES SOUND like something related (about one month later).

http://jfk14thday.com/jfk-gromyko-cold-war-1963/ brings more interesting news:

> For whatever reason, JFK didn't tape his famous meeting with Soviet Foreign Minister Andrei Gromyko on October 18, 1962, during the Cuban Missile Crisis. But nearly a year later, he did tape another Oval Office meeting with Gromyko.

The meeting that has been taped did not seem to contain any challenging material, so that they probably make him tape the second meeting to guarantee we would not associate his meeting with the USSR person with his death, I now reckon.

It does sound like a CIA plot in this case.

http://www.history.com/this-day-in-history/cuban-missile-crisis let us know the following:

On October 28, Khrushchev announced his government's intent to dismantle and remove all offensive Soviet weapons in Cuba. With the airing of the public message on Radio Moscow, the USSR confirmed its willingness to proceed with the solution secretly proposed by the Americans the day before. In the afternoon, Soviet technicians began dismantling the missile sites, and the world stepped back from the brink of nuclear war. The Cuban Missile Crisis was effectively over. In November, Kennedy called off the blockade, and by the end of the year all the offensive missiles had left Cuba. Soon after, the United States quietly removed its missiles from Turkey.

They had the secret meeting as an argument to state that JFK had died because of the conversations with the USSR and the way he managed the missile crisis. On the other hand, we here see that the Americans should be pretty happy with his actions in October, the 28th, 1963.

Well, in this case, it is more likely that he died by the hands of the security guard and because of a personal reason.

In any hypothesis, the person who killed him would have caused major change in human kind: If the wife divorced him, the Kennedys would lose political and social prestige. If he remained alive and bothered dissipating rumours about his personal and sexual life, who would have died would be Bobby, it seems.

We now spy on the life of Bobby after JFK's death:

After the assassination of President Kennedy in 1963, Robert Kennedy continued to serve as Attorney General under **President Lyndon Johnson** until September 1964. That November, he was elected to the U.S. Senate to represent New York. As a senator, Kennedy spoke out against America's involvement in the **Vietnam War**. In March 1966, King applauded Kennedy's statement against the war, invoking the legacy of his brother, the former president: "Your great brother, carried us far in new directions with his concept of a world of diversity; your position advances us to the next step which requires us to reach the political maturity to recognize and relate to all elements produced by the contemporary colonial revolutions" (King, 2 March 1966). The following year, King delivered his most comprehensive speech on the war, **"Beyond Vietnam"** to a crowd of over 3,000 people at Riverside Church in New York.

While campaigning for the presidency on 4 April 1968, Kennedy learned of **King's assassination** during a speech at a rally in Indianapolis, Indiana. Kennedy informed the largely black audience of King's death, cautioning them not to be "filled with hatred and distrust at the injustice of such an act, against all white people," for "Martin Luther King dedicated his life to love and to justice for his fellow human beings, and he died because of that effort" (Kennedy, 4 April 1968). Just two months later, Kennedy was assassinated in California while campaigning for the presidency.

The extract came from http://kingencyclopedia.stanford.edu/encyclopedia/encycloped

ia/enc_kennedy_robert_francis_
19251968/

Oh, so he was also assassinated, this in 1968, but he became a senator for the US following JFK's death, in 1964.

And we here see again the move of trying to associate his death with racist tensions.

We understand his own wife did not want to be with him anymore, but people would not let her divorce, what would easily give us a reason for her to kill the husband.

If he became the president, so much worse….

It could be then that the cause for all were actually Ethel.

See a paragraph that came from https://www.maryferrell.org/pages/Robert_Kennedy_Assassination.html:

In the assassinations of President Kennedy and Martin Luther King, Jr., the evidence tying the alleged assassins to the case was circumstantial and almost too neat. But here, Sirhan was apprehended on the scene firing a gun within a couple of feet of Kennedy. An open-and-shut case? Ironically, the RFK assassination has the starkest physical and eyewitness evidence indicating a conspiracy involving Sirhan and at least one additional gunman.

A few guns killed Bobby, not just one.

In any hypothesis, just like with JFK, all suspects are males.

They were both shot and only one bullet was fatal in both cases.

In any hypothesis, regardless of it being a plot of Ethel or those men acting on their own, a single person killed each one of them.

The power of one: Taking the place of God and judging others, taking the place of our democratic institutions and judging others, and then condemning to the penalty of having their lives subtracted from them.

Why not Institutions?

In 2001, we ourselves trusted the government of Australia, just to end up suffering the worst world atrocities for now 16 + years.

Somewhere between 1971 and 1991 Maristela Perozzo and family trusted the government of Brazil and the institution Army. That was just a Physical Education class and Olinto, her son, was attending it.

He had to run with everyone else for a certain amount of time. At a certain height, he turned to the instructor and said that he could not do it anymore, that he was unwell. The instructor responded by telling him that he could. He said it again and again, and the instructor just repeated his response.

He then died from a heart attack during that exercise.

I met a cop in Rio de Janeiro whilst I was in the waiting room of the military club, dentist, in the CBD. I and everyone who was there, in that waiting room with me, heard the guy bragging about the fact that he was the one who killed the 12 Candelaria minors (all got shot to death, episode that attracted the United Nations back then) but who was arrested was his fellow, who was actually innocent.

In this case, even the United Nations failed as an institution.

I still remember the carioca press trying to pretend to give a word to the guy, and he just said that it was not him.

Police, press, government, army, and all else: All failed. That man probably died in jail. The actual assassin is probably still bragging.

Soon before coming to Australia, somewhere between 1998 and 1999, I got the doorman of SENAI/CETIQT calling me for a conversation, but he had never done that.

He had been threatened. A boy threatened his life as a result of his actions, which were perfectly compatible with his function.

He said that next time he saw the boy the bullet that the boy dropped in his hand would be in his body.

I rushed and told Anna Fillipecki, the coordinator, and I then insisted in telling someone above her after she said that the man was nothing, that I should not care about that, just a doorman.

She actually threatened my own safety and welfare as I insisted in speaking to someone above her and she actually went with me, just like a guard would, all the way to the room of Neander, the person right above her in the vertical scale. I wanted to speak to Miran Esteves, the only one I would trust for these matters, but she said that was the maximum she was going to allow and she was threatening me in a really scary way. She had never done that before. She never acted like that with me before that day.

Neander was made aware. We could now consider that SENAI/CETIQT was aware, is it not?

The man got shot to death in front of my students in the following weekend. He died at the atrium with a few watching his death.

Claudio Perozzo died from being shot and was found inside of his car.

Witnesses saw a blonde woman with him in the car shortly before his death.

He was told to have gone with his wife, Carmen, to pick up a piece of clothing in the slums in his car.

She had reasons to do it: He was about to ask to divorce her and he managed to get proofs against her as to state she betrayed him, what meant she would take no money from the split.

Maristela Perozzo died without her going to jail and for suffering threats herself due to her persistence in terms of chasing justice for her son, Claudio.

Police, and justice failed, at least those.

Who would not trust Tom Cruise?

http://humanrightseverywhere. blogspot.com.au/2016/04/is- tom-cruise-confessing-crime- on-air.html brings something about a guy who decided to do rock climbing with Tom Cruise.

The own Tom Cruise seems to be saying they deliberately diminished the amount of Oxygen the guy would be getting and the guy effectively almost died because of that.

See: https://www.youtube.com/wat ch?v=Ecwh7g5GnP0

Why do I say that Tom Cruise equates institution and not individual?

Tom Cruise is trusted because of his marketing image, not because of himself as an individual most of the time.

The person who went with him in this climbing trip was more than likely someone who knew him by his commercial name, Tom Cruise. It is possible that his friends know him by his advertised non-commercial name, Thomas Cruise Mapother IV (http://www.wargs.com/other/mapother.html), but it is also possible that his actual name is completely different from that.

Tom Cruise is something like a brand, say IBM, to most of us.

We trust the brainwashing we receive about his name and person and go with it, most of the time.

I think I am sure the guy who almost lost his life is one of us, one of the members of the mass.

It is possible that he acquired a few physical problems despite surviving that trip. One of the possible problems has to do with his mental faculties.

We don't trust institutions because they frequently fail. Simple scientific reasoning tells us that that is stupid: We would all think that every institution has at least one member that does the opposite to all that is expected in the place. We also would all think that there is a chance we will get that one when dealing with the institution on a random day. We all know that anyone can change moral choices and behaviour from one second to another, and it is way more likely that they get worse than that they get better, so that even those that we think we know might become the one that cannot be trusted from one second to another. Even if we assume that all the members of a certain institution are absolutely moral, righteous, etc., there is a chance they get bugged in their heads and therefore criminally forced to become what they are not, so say the opposite to all that is expected, and even to all that they have presented that far in life. That can happen at any time.

An ultramarathon in Australia, in September, 2011, caused a few deformities in a few people. See:
http://www.watoday.com.au/w

a-news/inquiry-hands-down-
damning-findings-into-
kimberley-ultramarathon-
20120816-24ain.html

These people would have trusted the government of Australia, the organization that planned the marathon, the tradition of marathons in Australia, and whatever else that were an institution.

We must trust individuals if we want to have more chances of winning.

The Machine

Some people believe in the existence of a thing called Machine.

There is a TV series called Person of Interest (http://www.imdb.com/title/tt 1839578/) where the main actors make use of a machine to predict and stop crime.

Usually the term is seen associated with The State however. We then refer to how our lives and beings are manipulated without awareness or consent by the mechanisms of

The State, so say dominant media, official systems and organs, and things like that.

It seems that all systems and mechanisms in place can be used to make us believe something to an extreme, so say that the reporters have so much passion for what they do, and so much good intention, that they frequently give their lives in exchange for making us aware of something important, so say human rights violations.

We receive this message in multiple ways: when we are little, we have to associate pictures of men holding microphones with news, when we watch TV, the own news, we are told that a few more reporters have lost their lives whilst trying to broadcast from a region where they have wars, when we watch TV, some series, we are told that reporters may be called Clark Kent, and spend their lives doing good, saving the oppressed, punishing the evil doers, and so on. When we go to the movies, we may be told the same thing again: They may be called Clark Kent, wear their superman costume under their clothes, and, upon being told that someone is in danger, rush to save them from their previously set to happen criminal death.

As if it did not suffice, we open a magazine at our dentist's office, and see the ad displaying the image of Clark Kent in his Superman costume.

Then we think we have said enough, but there is still pencil case with Superman all over it, pencil sharpener, and so on.

Perhaps our mother will buy some crosswords and there we will see: Alternative name for Clark Kent, TV series.

That is The Machine.

Their mechanisms are never ending, since we still have costumes for sale in main department stores, really popular ones, leaflet distributors who dress just like Superman, and so on.

This is done in such a way that everyone on earth ends up believing reporters in general are really good people, people that can be trusted, and people that we can count on.

Once a figure enters The Machine, it will stick to our brain, it will become part of our Inner Reality (please consult our work on the human psyche), and we will incorporate all that junk that came from it into our personal universe without any criticism or sifting.

The Machine has the power of extending our ID: Perhaps we will see someone dressed as Superman when a bomb is coming toward us and we will then hold that person hoping that their structure will save our lives.

We think of Tom Cruise and we then think perfect husband, absolutely romantic and faithful, brilliant man, of wonderful and humanitarian ideas, clean actor, no drugs, completely healthy, if we grow up in Brazil between 1980 and 1999. Tom Cruise entered The Machine and we swallowed all that junk without thinking and without noticing that that was what we were actually doing.

The police, justice, the government, the press, and so on: All entered The Machine in most democratic countries.

What that means is that we created internal push buttons for every occasion they brainwashed us with: If you are in danger, call 000. If you have an issue with your own mother that can't be resolved in an amicable way, try Judge Judy.

Oh, well, some will call 000 because the instruction became part of their IDs. They will then be told to speak louder, and to tell exactly where they are, despite the variety of electronic resources available to the police nowadays, which would have to include localization via mobile and landline. Because they called 000 and the operator made them do that, the criminal found and killed them. This will never be told on air, different from the occasions where it all worked as advertised.

Because the person decided for Judge Judy instead of conversing with a third party, the mother actually had a heart attack and died.

The mother was right in all and if she ever made to The Judge Judy Show, the own Judge Judy would have told her that.

That, once more, will never be told to us. That is what we could perhaps call anti-Machine statistics.

The Machine is then a very dangerous thing: It can make us extend our ID to the level of automating our reactions to certain actions.

Before The Machine acts on us, our reactive system may actually be the right one for us.

We won't notice how or when, but that particular instruction will be there thanks not only to subliminal, as we have here said.

We start to understand that anyone, anything, and at any time, can be changed into something good and wanted, something that we will automatically invite to our lives, to our intimacy even, like it suffices they enter The Machine.

That makes us notice that the who and/or the what is

irrelevant and it is all perhaps money-dependant.

It is all very unfortunate, since even though Capitalism preaches equality through access and opportunity, we would like to think that those who enter The Machine really deserve being there because of moral and/or welfare reasons.

We have just mentioned the reasons as to why it is about money without making it explicit: Equality via access and opportunity, so equality of access and opportunity also in terms of entering The Machine.

That means that whoever has the money will dictate how we live and are either in part or in full.

That does agree with the thinking of a few very famous philosophers, such as Plato.

The quote you are about to read came from http://philosophyisnotonlyforsophie.blogspot.com.au/2015/06/justice.html

Each ruling class makes laws that are in its own interest, a democracy democratic laws, a tyranny tyrannical ones and so on; and in making these laws they define as 'right' for their subjects what is in the interest of themselves, the rulers, and if anyone breaks their laws he is punished as a *wrongdoer*. That is what I mean when I say that *right* is the same thing in all states, namely the interest of the established ruling class, and this ruling class is the *strongest* element in each state, and so if we argue correctly we see that *right* is always the same, the interest of the strongest party (Plato, Republic, p. 66, 338e-339a).

Plato was referring to the Oppressed x Oppressor Theory, a theory that I heard existed when speaking to really heavy marginals, such as Renato Gaui Filho, Lea Ricci Pinheiro, and Anna Fillipecki. These are modern marginals, the so called white collars: People who will spend their entire lives committing crimes, therefore doing wrong, but will never be recorded in the official books as having done that, and therefore will never be punished either.

The defenders of such a theory will say that if one chooses to defend the oppressed when their natural side is Oppressor, they will end up with them, that is, they will become Oppressed.

It is obviously easy to see that one could not inhabit both worlds: If one is to the side of The Oppressors, one cannot have scruples, for anything can go in The Machine. If one is to the side of The Oppressed, one is committed to defending, for instance, loyalty in competition, say the Country we talk about is Australia or Brazil, and this will make The Machine become heinous, and therefore will cause its destruction.

Plenty will defend that The Machine cannot be destroyed: We all love TV, movies, magazines, newspapers, and, even if we don't like those, there will always be something that can be called Machine, so say random letters printed on a T-shirt, poems given in Xmas cards, boards and signs that bring catching sentences, and things like that.

If The Machine cannot be destroyed, and nobody can be both Oppressor and Oppressed, the only way out would be creating an alternative team, so say the Non-Platonic.

The Non-Platonic, just to try to start something, could be those who believe part of what The Oppressed preach and part of what The Oppressed defend, so say they agree that The Machine will always exist, but they think that we can create filters to program its use, and, with this, nobody will be brainwashed with anything that be not a universal agreement, something considered good.

The Non-Platonic will then say that everyone would have to call 000 if they have a heart attack, so that we can brainwash people with that, and therefore that can go in The Machine. On the other hand, since people are actually dying due to the questions asked, we could have a code for heart attack and a device to locate the person when they dial.

If we follow the Non-Platonic, we can now be something else, not only Oppressed and Oppressor, but also those who defend criteria and selection in terms of what goes in The Machine, and criteria and selection that go beyond money.

The interesting thing here is that we can go through Priest's book on Nonclassical Systems, a book from 2001, to open our minds and search for new possibilities, new groups. This book is supposed to be a collection of all the Nonclassical Systems,

systems that are logical, in existence in 2000. These systems were created by means of perverting a few rules from Classical Logic. Classical Logic brought Cartesian thinking to the plate, that is, the thinking that everything was either false or true, nothing else. With Classical Logic, there is nothing beyond that, and that looks a lot like the Platonic thinking we here exposed.

The Machine is so dangerous that we could actually use a rule that perhaps derives from Supervaluationism: Only what is considered true in all evaluations should persist.

Call 000 when having a heart attack could be one of those.

In being a Non-Platonic person, however, we could now say that we first need to fix the system: All call centres of the type 000 should be equipped in a way to allow them to locate the place from where the call is made and to identify the situation of heart attack by means of a code, say an extra 0 in the keypad.

The Power Given to One

How can the power be given to one?

It seems that whatever passes through The Machine becomes plenty, becomes an institution, just like Tom Cruise.

We need to make people believe that there is only macro and micro around them, not anything else, so that they think that whatever happened when they were little in their family nest is simply a smaller instance of what happens to them when they are married and at the age of 30 y.o.: The problems are the same, especially the most fundamental ones.

Human life is a repetition of cycles, schemes, trades, etc., very much like the life of the animal.

One of the things that come as a plus in terms of human kind is the amount and nature of those.

There is always a micro and a macro situation, like for almost everything or for everything.

You are little and you are negotiating with your mother: Mum, let me go out to play. She says: No, it is too late. You have to wake up early for your school. You say: Mum, I will prove to you that I can do that if you let me stay out for a bit longer.

You may win and you may lose: If you win, you got something you really wanted, and, with it, a responsibility, which is that of delivering what you promised (I will prove to you that I can manage on my own). If you lose, you will once more just obey, and things will work in the same way they have always worked.

You are the CEO of IBM and you say: Boss, let us expand our company to Alaska. The boss says: No, there is literally nobody there. You say: Boss, I will prove to you that I can manage and we will still profit.

If you win, but you don't show the boss that you now have more profit, you may never get a job as a CEO again, like during the term of your life. If you lose, nothing has changed, and all will work as always.

Micro was your universe when you were a kid: You and your mother. Macro is your universe when you are employed by IBM as a CEO: The entire brand, which is worldly famous, depends on you, and, with it, gazillions of human beings.

If you can do things in a successful manner when you are little and have the situation of playing outside for longer, you can probably solve the situation involving IBM and Alaska because the skills are very similar if not equal.

Tom Cruise is a distant uncle who everyone talks about, but you have never met in person. Everyone says he is fantastic: a truly admirable man, completely moral, clean, healthy, socially useful, helpful, etc. You may meet him and get all disappointed, if you have expectations. You may never meet him. Repeating who he is without knowing him yourself is increasing the power of The Machined Tom Cruise.

Perhaps it is best ignoring what others say, keeping that as a very distant token in your memory. This until you personally meet and assess Tom Cruise.

The Bible says that we should test people before we classify them as One of Us. That sounds like the best advice ever given.

Perhaps it is always the case that knowing a person well can pay more than knowing thousands of people superficially, since those you know you can really count on for certain things. Even if they get bugged in their heads to have their psyche completely modified, the fact that you know them so well will make you be able to find out that that is what actually happened to them, and that has to give you strategic advantage.

The more people you know well, the wealthier you are. People talk a lot about money, but knowing others does mean money: You need someone to keep your car whilst you travel, and if you leave it with the wrong people, it may not be there anymore when you come back. That is actual money!

You need to treat your teeth. The wrong dentist may extract them all and now you have no natural teeth left and you had to pay for all that plus denture. Now you cannot effectively chew anything because the denture is bad and the service was bad, but, even if they were OK, the denture and the service in terms of the denture, people who have dentures do go through way more sacrifices when chewing. The right dentist might treat them all, keep them as natural, and charge the same price.

You ask me, someone who would perhaps be classified as a Maven by the author of Tipping Point, and I will tell you about the best dentist I have ever visited, since I inherited that concern from my parents. That dentist will indeed be the best available or one of those.

If you know me well, then you know you can trust me *for those*. You are in a TV show, the question is worth millions, and you are allowed to call a friend. You call me, and the prize is yours, basically.

Knowing is money, but knowing people will guarantee that your knowing is actually money, since you may know all but lose all that you know and all that you could have achieved with all that you know because you simply did not know somebody who could help you protect your knowledge or even your person against violence.

Horadam wanted me to put weight on my edges, the edges of my Starant Graphs, in 2002. I was working on disease spread and nothing that she was saying at that very moment connected to that, so that I could not relate to that, but that idea may serve us here: We are the centre of the Starant Graph, as you can see in my work, and all those ends, the nodes, would be the people who we know in this situation we here refer to. The darker those edges are, or the heavier, the more we know them. The darker the entire picture of our Starant is then, the more likely we are to succeed in life.

The darker the Starants of the nodes that connect to us are, the more likely we are to succeed in life as well: It is just that we start getting steps of separation between us and what is needed, say the information about the dentist.

We should aim at building a really heavy Starant Graph for ourselves.

Perhaps we could all have the picture of a Starant Graph on our walls and celebrate each time we add one node to it and each time an edge becomes heavier or darker.

That could work as a mandala for us: A geometric representation of our physical luck, the share of our luck that we can modify using our basic powers.

It is very unfortunate, but it is best if our nodes do not know each other: We actually have more power and more chances of having luck with our configuration if they don't.

Everyone we meet is someone of importance: Even the doorman may have valuable information to give to us someday.

Someone might know Tom Cruise and he might never help that person precisely because they know him, so say he knows them so well that he doesn't actually like them.

You, on the other hand, know the doorman and he will help you in the same situation, what will make you have what the other person who knew Tom

Cruise wanted. Who is wealthy and who is of importance now?

The importance of an individual in your Starant is given by how much they can contribute to your existence in a positive way.

If you always keep your mandala visible, you can also mark when you lose a node. If that happens, it is like going backwards several squares in a board game. You now must assess why that happened and then make sure you don't lose any other node for the same reason.

A successful person will have an ever-increasing mandala. A

person who goes through ups and downs will have an oscillating one: nodes that disappear, edges that weaken up, etc.

I can actually devise a world where people go around wearing their mandala and show it to the person they would like to marry as a condition of being accepted.

The mandala becomes our social power somehow.

Because it is money that everyone can get, it is a truly democratic measure of greatness.

Besides, that does mean social utility, love and appreciation of others, therefore positive impact on society, etc.

Conclusion

The power of one, now represented also in the shape of a Personal Mandala, represents ALSO your power.

Notice that the Starant Graph representing your circle of acquaintances and yourself may easily be folded to give you a flower: The more petals, the more beautiful and stronger it is.

Each petal that disappears is a loss of no dimension and, if that flower is detached, that means that you will not be able to replace it.

Your mandala may serve the purpose of making you control how you are doing in terms of wealth and investment: The more you are able to keep the already existing edges and nodes and strengthening those, the more wealth you should have. At the same time, you need to invest, to add edges and nodes. Just like everywhere else in human life, you need to adopt an aggressive attitude once in a while, even though part of your existence may be something like a guardian, just watching over what has already been achieved.

Institutions may translate even into Tom Cruise: They are abstract things.

As abstract entities, the likelihood of failure is really impressive.

It is definitely worth it concentrating on individuals, on the ones, instead of on those.

Plato and a few modern thinkers seem to believe the world is made of Oppressed and Oppressors, but there are other possible teams.

It is true that if we only have those teams, there is no possible intersection, that is, nobody can be part of both Oppressors and Oppressed.

That is because The Machine is part of The Oppressors and whoever is with The Oppressed would be committed to destroying it.

Destroying The Machine is not something that is possible, since there will always be a way to identify it in human life, like it is always present.

The Machine makes us even add instructions to our ID, that is, makes us extend it without noticing it.

The book about Nonclassical Logic, book published in 2001 by Priest, a collection of Nonclassical Systems, should help us open our minds and create beautiful alternatives to the Platonic World: More groups are viable.

One of those alternatives is clearly having The Machine as something that is used exclusively when some basic criteria are satisfied.

We should definitely make all our systems and mechanisms emphasize the power of one, what means changing all we have been doing to precisely the opposite.

All that really matters is the individual and their so specific set of skills, qualities, and so on.

The rules are a good thing but they tend to change it all into the power of many once more.

We need to work more on the power of the one, the power of the individual: Lots of thinking and labour have to be put into this Herculean task.

As said on our work about Synthetisers and Specifiers, it is perhaps only a DNA number that can mark us as individuals in a universal way.

Perhaps we should start referring to ourselves in such a way.

The more our discourse means individual, the power of one, the more we are approaching our target.

We really need to do things in a holistic way, so that psycholinguistics is also very important.

When power is fully given to one, our problems with injustice, crime, cowardice, disloyalty, parasitism, and others may actually start disappearing.

Nobody is an institution: We are always individuals.

Modern psychologists have said that we live in a schizoid world these days because we preach one thing and do another.

It may as well be true also here.

We notice that the Hollywoodian movies, for instance, do emphasize the power of one: Mission Impossible, Superman, Sixth Sense, etc.

Yet, at the same time, they put the institution to mean that person: the press, the police, the justice system, etc.

We have to realistically detach: We are not institutions and they represent us at most, we do not represent them, we represent ourselves.

Sometimes we agree with what they preach, so that we may say that we are the institution, but that is still equivocated, since the institution is usually its top management in first place, and therefore another set of individuals which may as well be regarded as one individual in our Inner Reality.

To teach people how to get to the power of one, we can teach them how to study macro situations from going through micro ones.

All in human life seems to be repetition, just like in the animal life.

Human life is a repetition of cycles, schemes, trades, etc.

One of the things that come as a plus in terms of human kind is the amount and nature of those.